John Bredakis

The Zeta function for those outside the top club of Prime Numbers Theorem

GRIN Verlag

Bibliografische Information der Deutschen Nationalbibliothek:

Die Deutsche Bibliothek verzeichnet diese Publikation in der Deutschen National-
bibliografie; detaillierte bibliografische Daten sind im Internet über http://dnb.d-
nb.de/ abrufbar.

Impressum:

Copyright © 2012 GRIN Verlag GmbH
Druck und Bindung: Books on Demand GmbH, Norderstedt Germany
ISBN: 978-3-656-32087-6

Dieses Buch bei GRIN:

http://www.grin.com/de/e-book/204593/the-zeta-function-for-those-outside-the-
top-club-of-prime-numbers-theorem

GRIN - Your knowledge has value

Der GRIN Verlag publiziert seit 1998 wissenschaftliche Arbeiten von Studenten, Hochschullehrern und anderen Akademikern als eBook und gedrucktes Buch. Die Verlagswebsite www.grin.com ist die ideale Plattform zur Veröffentlichung von Hausarbeiten, Abschlussarbeiten, wissenschaftlichen Aufsätzen, Dissertationen und Fachbüchern.

Besuchen Sie uns im Internet:

http://www.grin.com/

http://www.facebook.com/grincom

http://www.twitter.com/grin_com

The Zeta function for those outside the top club

of **Prime Numbers Theorem**

Realizing that the study of Zeta function ζ(s) is dependent on the Gamma

function Γ(s) , a function that I know well for s=x ε R , I decided to search

for the Zeta function in the internet ie: to get an overall satisfactory idea

about the Zeta function ζ(s) s=(σ+i.t).

Analytic everywhere except at the single pole at s=1 , with residue one.

Fortunately the Gamma function is analytic in the complex plane with certain simple poles.

For example: lim s->0 s.Γ(s) = lim s->0 Γ(s+1) = 0! =1   Residue one

In my effort to search for ζ(s) I had to overcome my shalow knowledge

of complex analysis and my ignorance of the various delicate methods of

numerical analysis. (Intense study of very high mathematics , of very long duration).

As I mentioned my purpose was to get a satisfactory over all idea of the Zeta function ζ(s) , without getting lost in the details of vast number of difficult to understand formulas.

From the initial frustration and dissapointment to the point of regretting my significant endeavour in higher mathematics,of very long duration ,to a final fullfilment of my purpose.   **At least I believe so** !!

I relied mainly on the following sources:

1. Wikipedia the free encyclopedia   2. An indroduction to the Riemann hypothesis by Theodore Yoder and 3. The zeros on the critical line of the Zeta function by Lorenzo Menici.

http// Mathhighways.blogspot.com / John Bredakis

An Introduction to the Riemann Hypothesis
And the role of the Gamma function

- For details see the pdf of Theodore J Yoder -

$$\theta(x) = \sum_{n=-\infty}^{n=+\infty} e^{-n^2 . \pi . x} \qquad \frac{\theta(x)-1}{2} = \sum_{n=1}^{+\infty} e^{-n^2 . \pi . x} \qquad \frac{\theta(x)}{\theta(1/x)} = \frac{1}{\sqrt{x}}$$

$$X(s) = \int_0^{+\infty} x^{s/2 - 1} . \left(\frac{\theta(x)-1}{2}\right) . dx = \pi^{-s/2} . \Gamma(\frac{s}{2}) . \zeta(s)$$

$$X(s) = \frac{-1}{s.(1-s)} + \int_1^{+\infty} \left[x^{s/2 - 1} + x^{(1-s)/2 - 1} \right] . \left(\frac{\theta(x)-1}{2}\right) . dx$$

$$\text{Real } s > 1$$

$$X(s) = \pi^{-s/2} . \Gamma(\frac{s}{2}) . \zeta(s) = X(1-s) = \pi^{-(1-s)/2} . \Gamma(\frac{1-s}{2}) . \zeta(1-s)$$

Equivalent
formula $\qquad \zeta(s) = (2.\pi)^{s-1} . 2 . \sin(\frac{\pi.s}{2}) . \Gamma(1-s) . \zeta(1-s)$

Analytic continuation - See text

$$\xi(s) = \frac{-s.(1-s)}{2} . X(s) = \frac{-s.(1-s)}{2} . \pi^{-s/2} . \Gamma(\frac{s}{2}) . \zeta(s)$$

$$\xi(s) = \frac{1}{2} - \frac{s.(1-s)}{2} . \int_1^{+\infty} \left[x^{s/2 - 1} + x^{(1-s)/2 - 1} \right] . \left(\frac{\theta(x)-1}{2}\right) . dx$$

Symmetric about the line Real s = (1/2)

$$\xi(\frac{1}{2} + i.t) = \frac{1}{2} - \frac{(1/4)+t^2}{2} . \int_1^{+\infty} x^{-3/4} . \cos\left[\ln(x) . \frac{t}{2}\right] . (\theta(x)-1) . dx$$

J.k.Bredakis MD

The first 42 zeros of the zeta function in the critical strip

14.134725142	1	$\zeta(1) \to +\infty$
21.022039639	2	This sum can be expressed as:
25.010857580	3	a product of prime numbers $\quad \zeta(3/2)=2.612$
30.424876126	4	(See Wikipedia)
32.935061588	5	$\zeta(2)=\pi^2/6 =1.645$
37.586178159	6	
40.918719012	7	$+\infty \quad s \qquad \zeta(3) \qquad =1.202$
43.327073281	8	$\zeta(s) = \Sigma \quad 1/n$
48.005150881	9	$n=1$
49.773832478	10	$\zeta(4)=\pi^4/90 =1.0823$
52.970321478		
56.446247697		
59.347044003		The zeta function in the complex plane has:
60.831778525		
65.112544048		
67.079810529		At s=1 a singularity:
69.546401711		(A single pole with residue one)
72.067157674		
75.704690699		At s=0 the value of $-(1/2)$
77.144840069	20	
79.337375020		
82.910380854		At s=-1 the value of $-(1/12)$
84.735492981		
87.425274613		
88.809111208		Trivial zeros at s = -2 , -4 , -6 , etc
92.491899271		
94.651344041		
95.870634228		The beauty of the zeta function is:
98.831194218		in the critical strip
101.317851006	30	0< Real s <1
103.725538040		with plenty of zeros
105.446623052		
107.168611184		The conjecture by Riemann is that:
111.029535543		all those zeros are in the line
111.874659177		Real s = (1/2)
114.320220915		
116.226680321		Notice that: $\xi(1/2 + i.t) = \xi(1/2 - i.t)$
118.790782866		
121.370125002		* Needless to say that the best
122.946829294	40	mathematical brains were
124.256818554	41	summond to develop
127.516683880	42	those supercomputer numbers

With the aid of the pdf by Theodore J Yoder at least iv got the idea of whats going on with zeta function.

- See the pdf by Jerome Baltzersen —

Hardy's theorem and the prime numbers theorem

Roots of the Zeta function ζ(s)

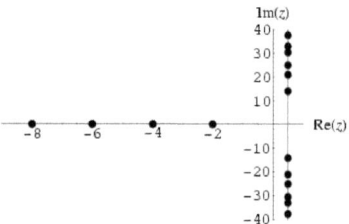

Relation between the Zeta function and the primes

$$\zeta(s) = 1 + \frac{1}{2^s} + \frac{1}{3^s} + \frac{1}{4^s} + \frac{1}{5^s} + \cdots$$

$$\frac{1}{2^s}\zeta(s) = \frac{1}{2^s} + \frac{1}{4^s} + \frac{1}{6^s} + \frac{1}{8^s} + \frac{1}{10^s} + \cdots$$

Subtracting the second from the first we remove all elements that have a factor of 2:

$$\left(1 - \frac{1}{2^s}\right)\zeta(s) = 1 + \frac{1}{3^s} + \frac{1}{5^s} + \frac{1}{7^s} + \frac{1}{9^s} + \frac{1}{11^s} + \frac{1}{13^s} + \cdots$$

Repeating for the next term:

$$\frac{1}{3^s}\left(1 - \frac{1}{2^s}\right)\zeta(s) = \frac{1}{3^s} + \frac{1}{9^s} + \frac{1}{15^s} + \frac{1}{21^s} + \frac{1}{27^s} + \frac{1}{33^s} + \cdots$$

Subtracting again we get:

$$\left(1 - \frac{1}{3^s}\right)\left(1 - \frac{1}{2^s}\right)\zeta(s) = 1 + \frac{1}{5^s} + \frac{1}{7^s} + \frac{1}{11^s} + \frac{1}{13^s} + \frac{1}{17^s} + \cdots$$

where all elements having a factor of 3 or 2 (or both) are removed.

It can be seen that the right side is being sieved. Repeating infinitely we get:

$$\cdots \left(1 - \frac{1}{11^s}\right)\left(1 - \frac{1}{7^s}\right)\left(1 - \frac{1}{5^s}\right)\left(1 - \frac{1}{3^s}\right)\left(1 - \frac{1}{2^s}\right)\zeta(s) = 1$$

Dividing both sides by everything but the ζ(s) we obtain:

$$\zeta(s) = \frac{1}{\left(1 - \frac{1}{2^s}\right)\left(1 - \frac{1}{3^s}\right)\left(1 - \frac{1}{5^s}\right)\left(1 - \frac{1}{7^s}\right)\left(1 - \frac{1}{11^s}\right)\cdots}$$

$$\theta(x) = \sum_{n=-\infty}^{n=+\infty} e^{-n^2.\pi.x} \qquad \frac{\theta(x)-1}{2} = \sum_{n=1}^{+\infty} e^{-n^2.\pi.x} \qquad \frac{\theta(x)}{\theta(1/x)} = \frac{1}{\sqrt{x}}$$

$$X(s) = \int_0^{+\infty} x^{s/2-1}.\left(\frac{\theta(x)-1}{2}\right).dx = \pi^{-s/2}.\Gamma\left(\frac{s}{2}\right).\zeta(s)$$

$$X(s) = \frac{-1}{s.(1-s)} + \int_1^{+\infty} \left[x^{s/2-1} + x^{(1-s)/2-1}\right].\left(\frac{\theta(x)-1}{2}\right).dx$$
$$\text{Real } s > 1$$

$$X(s) = \pi^{-s/2}.\Gamma\left(\frac{s}{2}\right).\zeta(s) = X(1-s) = \pi^{-(1-s)/2}.\Gamma\left(\frac{1-s}{2}\right).\zeta(1-s)$$

Equivalent formula
$$\zeta(s) = (2.\pi)^{s-1}.2.\sin\left(\frac{\pi.s}{2}\right).\Gamma(1-s).\zeta(1-s)$$

Analytic continuation – See text

$$\xi(s) = \frac{-s.(1-s)}{2}.X(s) = \frac{-s.(1-s)}{2}.\pi^{-s/2}.\Gamma\left(\frac{s}{2}\right).\zeta(s)$$

$$\xi(s) = \frac{1}{2} - \frac{s.(1-s)}{2}.\int_1^{+\infty} \left[x^{s/2-1} + x^{(1-s)/2-1}\right].\left(\frac{\theta(x)-1}{2}\right).dx$$

$$\xi\left(\frac{1}{2} + i.t\right) = \frac{1}{2} - \frac{(1/4)+t^2}{2}.\int_1^{+\infty} x^{-3/4}.\cos\left[\ln(x).\frac{t}{2}\right].(\theta(x)-1).dx$$

$$\xi\left(\frac{1}{2}+it\right) = \frac{1}{2}\left[\frac{1}{2}+it\right]\left[-\frac{1}{2}+it\right]\Gamma\left(\frac{1}{4}+\frac{it}{2}\right)\pi^{-\frac{1}{4}-\frac{it}{2}}\zeta\left(\frac{1}{2}+it\right)$$

$$= -\frac{1}{2}\exp\left[\operatorname{Re}\log\Gamma\left(\frac{1}{4}+\frac{it}{2}\right)\right]\pi^{-\frac{1}{4}}\left(t^2+\frac{1}{4}\right)Z(t),$$

$\theta(0)=0 \quad \zeta(1/2)=-1.46$

$$Z(t) = e^{i\vartheta(t)}\zeta\left(\frac{1}{2}+it\right) = \exp\left[i\operatorname{Im}\log\Gamma\left(\frac{1}{4}+\frac{it}{2}\right)-i\frac{\log\pi}{2}t\right]\zeta\left(\frac{1}{2}+it\right)$$

Z(t) is the Riemann Siegel function $\qquad \vartheta(t) \sim \frac{t}{2}\log\frac{t}{2\pi} - \frac{t}{2} - \frac{\pi}{8} + \frac{1}{48t}$

$$\xi(\frac{1}{2} + i.t) = \frac{1}{2} - \frac{(1/4)+t^2}{2} . \int_1^{+\infty} x^{-3/4} . \cos\left[\ln(x) . \frac{t}{2}\right] . (\Theta(x)-1) . dx$$

The proof of the above

$$\chi(s) := \int_0^\infty x^{s/2-1}\left(\frac{\vartheta(x)-1}{2}\right)dx = \int_1^\infty x^{s/2-1}\left(\frac{\vartheta(x)-1}{2}\right)dx \qquad (6)$$
$$+ \int_1^\infty x^{-s/2-1}\left(\frac{\vartheta(1/x)-1}{2}\right)dx.$$

We defined $\chi(s)$ simply for easy reference. Equation (6) looks good; not only do the limits of integration match but also we can use (5) to find,

$$\frac{\vartheta(1/x)-1}{2} = \frac{\sqrt{x}\,\vartheta(x)-1}{2} = \sqrt{x}\left(\frac{\vartheta(x)-1}{2}\right) + \frac{\sqrt{x}-1}{2},$$

and then it follows that from (6),

$$\chi(s) = \int_1^\infty (x^{s/2-1} + x^{(1-s)/2-1})\left(\frac{\vartheta(x)-1}{2}\right)dx$$
$$+ \int_1^\infty \frac{1}{2}\left[x^{(1-s)/2-1} - x^{-s/2-1}\right]dx.$$

Finally, an integral that is easily integrable. For $\Re(s) > 1$, the second integral converges. We get

$$\int_1^\infty \frac{1}{2}\left[x^{(1-s)/2-1} - x^{-s/2-1}\right]dx = \left[\frac{x^{(1-s)/2}}{1-s} + \frac{x^{-s/2}}{s}\right]_1^\infty = \frac{-1}{s(1-s)},$$

so that,

$$\chi(s) = \Gamma(\frac{s}{2})\pi^{-s/2}\zeta(s) \qquad (7)$$
$$= \frac{-1}{s(1-s)} + \int_1^\infty (x^{s/2-1} + x^{(1-s)/2-1})\left(\frac{\vartheta(x)-1}{2}\right)dx,$$

for $\Re(s) > 1$ [5, 9].

$$\frac{s(1-s)}{2} = \frac{(1/2+it)(1/2-it)}{2} = \frac{1/4+t^2}{2}.$$

It is obviously real for all $t \in \mathbb{R}$. Likewise, the integrand with $\sigma = 1/2$ is

$$(x^{s/2-1} + x^{(1-s)/2-1})(\frac{\vartheta(x)-1}{2}) = (x^{-3/4}x^{it/2} + x^{-3/4}x^{-it/2})(\frac{\vartheta(x)-1}{2})$$
$$= x^{-3/4}\cos\left(\ln(x)\frac{t}{2}\right)(\vartheta(x)-1), \qquad (15)$$

The formula of X(s) is derived by the assumption that Real s>1

$$\theta(x) = \sum_{n=-\infty}^{n=+\infty} e^{-n^2 \cdot \pi \cdot x} \qquad \frac{\theta(x)-1}{2} = \sum_{n=1}^{+\infty} e^{-n^2 \cdot \pi \cdot x} \qquad \frac{\theta(x)}{\theta(1/x)} = \frac{1}{\sqrt{x}}$$

$$X(s) = \int_0^{+\infty} x^{s/2 - 1} \cdot \left(\frac{\theta(x)-1}{2}\right) . dx = \pi^{-s/2} . \Gamma(\frac{s}{2}) . \zeta(s)$$

$$X(s) = \frac{-1}{s.(1-s)} + \int_1^{+\infty} \left[x^{s/2 - 1} + x^{(1-s)/2 - 1} \right] . \left(\frac{\theta(x)-1}{2}\right) . dx$$
$$\text{Real s} > 1$$

Notice that: X(s) = X(1-s)

$$X(s) = \pi^{-s/2} . \Gamma(\frac{s}{2}) . \zeta(s) = X(1-s) = \pi^{-(1-s)/2} . \Gamma(\frac{1-s}{2}) . \zeta(1-s)$$

- -

Equivalent
formula $\qquad \zeta(s) = (2.\pi)^{s-1} . 2.\sin(\frac{\pi.s}{2}) . \Gamma(1-s) . \zeta(1-s)$

Meaning that the formula of X(s) is also valid for Real s<1

$$\zeta(s) = \frac{\pi^{s/2}}{\Gamma(\frac{s}{2})} . X(s) = \frac{\pi^{s/2}}{\Gamma(\frac{s}{2})} . \left[\frac{-1}{s.(1-s)} + \int_1^{+\infty} \left[x^{s/2} + x^{(1-s)/2} \right] . x^{-1} . \left(\frac{\theta(x)-1}{2}\right) . dx \right]$$
$$\text{A convergent integral}$$

The above formula provides analytic continuation of $\zeta(s)$

and provides also the value of $\zeta(s)$ at s=0 ie: $\zeta(0)=-1/2$

Y=Euler's constant
|
=0.577215665 $\qquad \frac{1}{\Gamma(s)} = e^{\gamma.s} . s . \prod_{n=1}^{+\infty} \left[1 + \frac{s}{n} \right] . e^{-s/n} \qquad \prod \text{ stands for products}$

By analytic continuation of $\zeta(s)$
except for the pole (s=1)

This formula is analytic also in the critical strip
$0<$ Real $s <1$

$$\zeta(s)=\frac{\pi^{s/2}}{\Gamma(\frac{s}{2})}.X(s)=\frac{\pi^{s/2}}{\Gamma(\frac{s}{2})}.\left[\frac{-1}{s.(1-s)} + \int_1^{+\infty}\left[x^{s/2}+x^{(1-s)/2}\right].x^{-1}.(\frac{\theta(x)-1}{2}).dx\right]$$

A convergent integral

Therefore the following formulas are also analytic in the critical strip

$$\xi(s) = \frac{-s.(1-s)}{2}.X(s) = \frac{-s.(1-s)}{2}.\pi^{-s/2}.\Gamma(\frac{s}{2}).\zeta(s)$$

$$\xi(s) = \frac{1}{2} - \frac{s.(1-s)}{2}.\int_1^{+\infty}\left[x^{s/2 - 1} + x^{(1-s)/2 - 1}\right].(\frac{\theta(x)-1}{2}).dx$$

$$\xi(\frac{1}{2} + i.t) = \frac{1}{2} - \frac{(1/4)+t^2}{2}.\int_1^{+\infty}x^{-3/4}.\cos\left[\ln(x).\frac{t}{2}\right].(\theta(x)-1).dx$$

The tricks are

1.The analytic continuation of $\zeta(s)$
except for the pole (s=1)
With $\zeta(0)=-1/2$

and

2.The convergence of the integral

$$\zeta(s)=\frac{\pi^{s/2}}{\Gamma(\frac{s}{2})}.X(s)=\frac{\pi^{s/2}}{\Gamma(\frac{s}{2})}.\left[\frac{-1}{s.(1-s)} + \int_1^{+\infty}\left[x^{s/2}+x^{(1-s)/2}\right].x^{-1}.(\frac{\theta(x)-1}{2}).dx\right]$$

A convergent integral

$$\theta(x) = \sum_{n=-\infty}^{n=+\infty} e^{-n^2.\pi.x} \quad\bigg|\quad \frac{\theta(x)-1}{2} = \sum_{n=1}^{+\infty} e^{-n^2.\pi.x} \quad\bigg|\quad \frac{\theta(x)}{\theta(1/x)} = \frac{1}{\sqrt{x}}$$

Notice also:

$$\zeta(s) = \frac{\pi^{s/2}}{\Gamma(\frac{s}{2})} \cdot X(s) = \frac{1}{\Gamma(s)} \cdot \int_0^{+\infty} \frac{1}{(e^t - 1)} \cdot t^{s-1} \cdot dt = \frac{1}{\Gamma(s)} \cdot \sum_{n=1}^{+\infty} \cdot \int_0^{+\infty} e^{-n.t} \cdot M$$

$$M = t^{s-1} \cdot dt$$

$\boxed{0 < \text{Real } s < 1}$ $= \frac{1}{\Gamma(s)} \cdot \int_0^{+\infty} \left[\frac{1}{(e^t - 1)} - \frac{1}{t} \right] \cdot t^{s-1} \cdot dt$ ***** pdf on zeta function **by Petersen**

$\boxed{-1 < \text{Real } s < 0}$ $= \frac{1}{\Gamma(s)} \cdot \int_0^{+\infty} \left[\frac{1}{(e^t - 1)} - \frac{1}{t} + \frac{1}{2} \right] \cdot t^{s-1} \cdot dt$ *****

The Zeta function in the complex plain

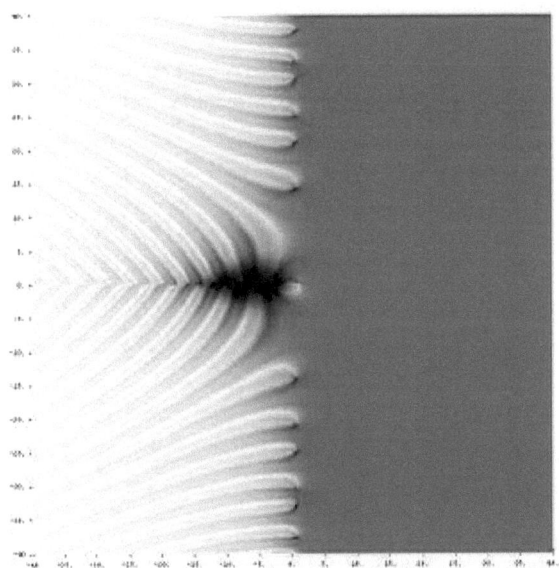

$$\xi(s) = \frac{-s.(1-s)}{2}.X(s) = \frac{-s.(1-s)}{2}.\pi^{-s/2}.\Gamma(\frac{s}{2}).\zeta(s)$$

$$\xi(\frac{1}{2} + i.t) = \frac{1}{2} - \frac{(1/4)+t^2}{2}.\int_{1}^{+\infty} x^{-3/4}.\cos\left[\ln(x).\frac{t}{2}\right].(\theta(x)-1).dx$$

$$\xi\left(\frac{1}{2}+it\right) = \frac{1}{2}\left[\frac{1}{2}+it\right]\left[-\frac{1}{2}+it\right]\Gamma\left(\frac{1}{4}+\frac{it}{2}\right)\pi^{-\frac{1}{4}-\frac{it}{2}}\zeta\left(\frac{1}{2}+it\right)$$

$$= -\frac{1}{2}\exp\left[\operatorname{Re}\log\Gamma\left(\frac{1}{4}+\frac{it}{2}\right)\right]\pi^{-\frac{1}{4}}\left(t^2+\frac{1}{4}\right)Z(t),$$

$$\theta(0)=0 \quad \zeta(1/2)=-1.46$$

$$Z(t) = e^{i\vartheta(t)}\zeta\left(\frac{1}{2}+it\right) = \exp\left[i\operatorname{Im}\log\Gamma\left(\frac{1}{4}+\frac{it}{2}\right)-i\frac{\log\pi}{2}t\right]\zeta\left(\frac{1}{2}+it\right)$$

Z(t) is the Riemann Siegel function
$$\vartheta(t) \sim \frac{t}{2}\log\frac{t}{2\pi} - \frac{t}{2} - \frac{\pi}{8} + \frac{1}{48t}$$

Odlyzko-Schönhage algorithm
A modification of the above Z(t)

$$Z(t) = 2\sum_{n\leq\sqrt{t/(2\pi)}}\frac{\cos[\vartheta(t)-t\log n]}{\sqrt{n}} + r(t) \qquad (2.34)$$

is approximately

$$r(t) \sim (-1)^{N-1}\left(\frac{t}{2\pi}\right)^{-\frac{1}{4}}\left[A_0 + A_1\left(\frac{t}{2\pi}\right)^{-\frac{1}{2}} + A_2\left(\frac{t}{2\pi}\right)^{-1} + A_3\left(\frac{t}{2\pi}\right)^{-\frac{3}{2}} + A_4\left(\frac{t}{2\pi}\right)^{-2}\right]$$

$$(2.35)$$

where $N = [\sqrt{t/(2\pi)}]$, $p = \{\sqrt{t/(2\pi)}\}$ *and*

$$A_0 = \Psi(p) = \frac{\cos[2\pi(p^2-p-\frac{1}{16})]}{\cos(2\pi p)},$$

$$A_1 = -\frac{1}{2^5\,3\pi^2}\Psi^{(3)}(p),$$

$$A_2 = \frac{1}{2^6\pi^2}\Psi^{(2)}(p) + \frac{1}{2^{11}\,3^2\pi^4}\Psi^{(6)}(p),$$

$$A_3 = -\frac{1}{2^6\pi^2}\Psi^{(1)}(p) - \frac{1}{2^8\,3\cdot5\pi^2}\Psi^{(5)}(p) - \frac{1}{2^{16}\,3^4\pi^6}\Psi^{(5)}(p),$$

$$A_4 = \frac{1}{2^7\pi^2}\Psi(p) + \frac{1}{2^{13}\,3\pi^4}\Psi^{(4)}(p) + \frac{1}{2^{17}\,3^2\cdot5\pi^6}\Psi^{(8)}(p) + \frac{1}{2^{23}\,3^5\pi^8}\Psi^{(12)}(p).$$

See also Haselgrove's tables

Haselgrove's table for $\zeta(1/2+i.t)$ in the range $0\leq t \leq 26.8$

t	$\zeta(1/2+i.t)$	t	$\zeta(1/2+i.t)$	t	$\zeta(1/2+i.t)$
0.0	-1.46	9.0	+1.45+0.19.i	18.0	+2.33-0.19.i
0.2	-1.18-0.67.i	9.2	+1.48+0.14.i	18.2	+2.27-0.45.i
0.4	-0.68-0.94.i	9.4	+1.51+0.08.i	18.4	+2.17-0.66.i
0.6	-0.28-0.94.i	9.6	+1.53+0.02.i	18.6	+2.02-0.86.i
0.8	-0.02-0.84.i	9.8	+1.54-0.04.i	18.8	+1.84-1.03.i
1.0	+0.14-0.72.i	10.0	+1.54-0.12.i	19.0	+1.62-1.16.i
1.2	+0.25-0.62.i	10.2	+1.54-0.19.i	19.2	+1.38-1.24.i
1.4	+0.32-0.52.i	10.4	+1.53-0.26.i	19.4	+1.13-1.28.i
1.6	+0.37-0.44.i	10.6	+1.50-0.34.i	19.6	+0.88-1.26.i
1.8	+0.41-0.37.i	10.8	+1.47-0.42.i	19.8	+0.65-1.18.i
2.0	+0.44-0.31.i	11.0	+1.42-0.49.i	20.0	+0.43-1.06.i
2.2	+0.46-0.26.i	11.2	+1.36-0.56.i	20.2	+0.25-0.90.i
2.4	+0.48-0.21.i	11.4	+1.29-0.62.i	20.4	+0.11-0.70.i
2.6	+0.50-0.16.i	11.6	+1.21-0.67.i	20.6	+0.02-0.48.i
2.8	+0.52-0.12.i	11.8	+1.12-0.71.i	20.8	-0.02-0.25.i
3.0	+0.53-0.08.i	12.0	+1.02-0.75.i	21.0	-0.01-0.02.i
3.2	+0.55-0.04.i	12.2	+0.91-0.76.i	21.2	+0.06+0.19.i
3.4	+0.56-0.01.i	12.4	+0.79-0.76.i	21.4	+0.18+0.35.i
3.6	+0.58+0.03.i	12.6	+0.68-0.75.i	21.6	+0.34+0.52.i
3.8	+0.59+0.06.i	12.8	+0.56-0.71.i	21.8	+0.52+0.62.i
4.0	+0.61+0.05.i	13.0	+0.44-0.06.i	22.0	+0.72+0.67.i
4.2	+0.62+0.12.i	13.2	+0.33-0.58.i	22.2	+0.92+0.66.i
4.4	+0.64+0.15.i	13.4	+0.23-0.49.i	22.4	+1.11+0.60.i
4.6	+0.66+0.18.i	13.6	+0.35-0.38.i	22.6	+1.26+0.49.i
4.8	+0.68+0.21.i	13.8	+0.07-0.24.i	22.8	+1.38+0.34.i
5.0	+0.70+0.23.i	14.0	+0.02-0.10.i	23.0	+1.45+0.16.i
5.2	+0.73+0.26.i	14.2	-0.01+0.04.1	23.2	+1.46-0.03.i
5.4	+0.75+0.28.i	14.4	-0.01+0.21.i	23.4	+1.41-0.21.i
5.6	+0.78+0.30.i	14.6	+0.01+0.38.i	23.6	+1.30-0.38.i
5.8	+0.81+0.32.i	14.8	+0.07+0.55.i	23.8	+1.14-0.50.i
6.0	+0.84+0.34.i	15.0	+0.15+0.70.i	24.0	+0.95-0.58.i
6.2	+0.87+0.36.i	15.2	+0.26+0.85.i	24.2	+0.73-0.60.i
6.4	+0.91+0.37.i	15.4	+0.39+0.98.i	24.4	+0.51-0.55.i
6.6	+0.94+0.38.i	15.6	+0.56+1.09.i	24.6	+0.30-0.43.i
6.8	+0.98+0.39.i	15.8	+0.74+1.17.1	24.8	+0.13-0.21.i
7.0	+1.02+0.40.i	16.0	+0.94+1.22.i	25.0	0.00-0.01.i
7.2	+1.06+0.40.i	16.2	+1.15+1.23.i	25.2	-0.05+0.26.i
7.4	+1.11+0.40.i	16.4	+1.36+1.20.i	25.4	-0.04+0.55.i
7.6	+1.15+0.39.i	16.6	+1.57+1.14.i	25.6	+0.06+0.85.i
7.8	+1.20+0.38.i	16.8	+1.77+1.04.i	25.8	+0.25+1.11.i
8.0	+1.24+0.36.i	17.0	+1.95+0.90.i	26.0	+0.50+1.34.i
8.2	+1.29+0.34.i	17.2	+2.10+0.72.i	26.2	+0.82+1.49.i
8.4	+1.33+0.31.i	17.4	+2.22+0.52.i	26.4	+1.17+1.56.i
8.6	+1.37+0.28.i	17.6	+2.30+0.29.i	26.8	+1.55+1.54.i
8.8	+1.41+0.24.i	17.8	+2.34+0.06.i	26.8	+1.92+1.42.i

Roots of $\zeta(1/2+i.t)$ in the range $0\leq t \leq 26.8$

14.134725142
21.022039639
25.010857580

An impressive formula - Lanczos approximation

For any complex argument with nonegative real part

Error $|\varepsilon| < 2 \times 10^{-10}$ for Real $z > 0$ (Rez>0)

$$\Gamma(z) = \left(\frac{\sqrt{2\pi}}{z}\left(p_0 + \sum_{n=1..6}\frac{p_n}{z+n}\right)\right)(z+5.5)^{z+0.5}\,e^{-(z+5.5)}$$

$p_0 = 1.000000000190015$
$p_1 = 76.18009172947146$
$p_2 = -86.50532032941677$
$p_3 = 24.01409824083091$
$p_4 = -1.231739572450155$
$p_5 = 1.208650973866179 \cdot 10^{-3}$
$p_6 = -5.395239384953 \cdot 10^{-6}$

$$\xi(s) = \frac{-s.(1-s)}{2}.X(s) = \frac{-s.(1-s)}{2}.\pi^{-s/2}.\Gamma(\frac{s}{2}).\zeta(s)$$

$$\xi(s) = \frac{1}{2} - \frac{s.(1-s)}{2}.\int_1^{+\infty}\left[x^{s/2-1} + x^{(1-s)/2-1}\right].(\frac{\theta(x)-1}{2}).dx$$

$$\xi(\frac{1}{2}+i.t) = \frac{1}{2} - \frac{(1/4)+t^2}{2}.\int_1^{+\infty}x^{-3/4}.\cos\left[\ln(x).\frac{t}{2}\right].(\theta(x)-1).dx$$

$$\xi\left(\frac{1}{2}+it\right) = \frac{1}{2}\left[\frac{1}{2}+it\right]\left[-\frac{1}{2}+it\right]\Gamma\left(\frac{1}{4}+\frac{it}{2}\right)\pi^{-\frac{1}{4}-\frac{it}{2}}\zeta\left(\frac{1}{2}+it\right)$$

$$= -\frac{1}{2}\exp\left[\operatorname{Re}\log\Gamma\left(\frac{1}{4}+\frac{it}{2}\right)\right]\pi^{-\frac{1}{4}}\left(t^2+\frac{1}{4}\right)Z(t),$$

$\theta(0)=0 \quad \zeta(1/2)=-1.46$

$$Z(t) = e^{i\vartheta(t)}\zeta\left(\frac{1}{2}+it\right) = \exp\left[i\operatorname{Im}\log\Gamma\left(\frac{1}{4}+\frac{it}{2}\right) - i\frac{\log\pi}{2}t\right]\zeta\left(\frac{1}{2}+it\right)$$

$$\vartheta(t) \sim \frac{t}{2}\log\frac{t}{2\pi} - \frac{t}{2} - \frac{\pi}{8} + \frac{1}{48t}$$

Z(t) is the Riemann Siegel function

- Definition of a natural logarithm of a complex number -

The logarithm can be used to define arbitrary powers, such as raising the complex number t to a complex power z

$$\ln(z) = \ln\left[|z|.e^{i.(\theta + 2k\pi)}\right] = \ln|z| + i.[\theta + 2k\pi] \quad \boxed{|z|=r}$$

$$|\ln(z)| = \sqrt{\ln^2(r) + (\theta+2k\pi)^2} \quad \text{Arg } \ln(z) = \arctan\left(\frac{\theta+2k\pi}{r}\right)$$

Both depend on the value of k

The domain is usually restricted to the principal value

$$\boxed{0\le \theta <2\pi} \quad \text{corresponding to } k=0$$
$$\text{(see above)}$$

$$w = t^z \Rightarrow \ln(w) = z.\ln(t) \quad \text{Suppose that } z = (x + i.y)$$

$$\ln(t) = \ln\left[|t|.e^{i.(\theta + 2k\pi)}\right] = \ln|t| + i.[\theta + 2k\pi] \quad \boxed{|t|=r}$$

$$z.\ln(t) = x.\ln|t| - y.(\theta+2\kappa\pi) + i.[x.(\theta+2\kappa\pi) + y.\ln|t|]$$

In conclusion:

$$\ln(w)=\ln(t^z)=\left[x.\ln|t|-y.(\theta+2k\pi) + i.[x.(\theta+2k\pi)+y.\ln|t|]\right]$$

$$z = (x + i.y) \quad \boxed{0\le \theta < 2\pi} \quad \begin{array}{l}\text{Iff Restricted}\\ \text{to principal} \quad (k=0)\\ \text{value}\end{array}$$

$$|\ln(t)| = \sqrt{\ln^2|t| + (\theta+2k\pi)^2} \quad \text{Arg } \ln(t) = \arctan\left(\frac{\theta+2k\pi}{|t|}\right)$$

Simplifying things

$$(a+i.b)^{(c+i.d)} = r^c.e^{-d.\theta}.[\cos(c.\theta + d.\ln r)+i.\sin(c.\theta + d.\ln r)]$$

$$r = \sqrt{a^2 + b^2} \quad \theta = \arctan(b/a) \quad \begin{array}{l}\text{Note: } \theta \text{ must be in the}\\ \text{correct quadrant!}\end{array}$$

Some additional remarks on the Zeta function

for my level of mathematics outsise this top club

An observation: In general:

$$\frac{10}{\ln 10} \sim 4 = \pi(10) = \boxed{\begin{array}{c}\text{Number of primes}\\\text{up to the number 10}\end{array}} \qquad \pi(x) \sim \frac{x}{\ln x}$$

$$2\ 3\ 5\ 7 \qquad\qquad\qquad\qquad\qquad \text{As } x \to +\infty$$

Bernoulli's numbers

$$B2 = \frac{1}{6}$$

$$\zeta(-n) = (-1)^n \cdot \frac{Bn+1}{n+1}$$

$$B4 = \frac{-1}{30}$$

$$\boxed{\begin{array}{c}\text{Odd Bernoulli's}\\\text{numbers}\\\text{with the exception}\\\text{of B1 vanish}\end{array}}$$

$$B6 = \frac{1}{42}$$

$$B8 = \frac{-1}{30} \qquad\qquad \zeta(2n) = \frac{(2\pi)^{2n} \cdot (-1)^{n+1} \cdot B2n}{2 \cdot (2n)!}$$

- -

$$\zeta(s) = \left[1 - \frac{2}{2^s}\right]^{-1} \cdot \sum_{n=1}^{+\infty} \frac{(-1)^{n+1}}{n^s}$$

Converges uniformly for Real s>0

See also the pdf by Theodore Yoder

for repulsion between the zeros of $\zeta(s)$

in the critical line

Quoted from Efthimios Kaxiras

On his excellent pdf on Practical Mathematics

Department of Physics and Applied Sciences

Harvard University

Mathematics is the language of science

Someone can learn a new language in a formal way and have
a deep knowledge of the many aspects of the language
                    (grammar , literature etc)
This takes a lot of effort and at least in the beginning
can be tiresome.
On the other hand one can pick up the essentials of a
language , without a deep knowledge of all the richness.
It is a bit like learning a language at a street - smart
level , without being exposed to its fine points.
            Of course much is missed in this way
But the satisfaction of being able to quickly and effecti-
vely use the language for new experiences may compensate
for this.

This is exactly what I feel
dealing with the Zeta function

References:

1. **Higher Mathematics for beginners:**
   ----
   by Ya.B.Zeldovich
   (Mir Publishers Moscow 1973)

2. **Calculus with analytic geometry:**
   ----
   by Harley Flanders and Justin J Price
   (Academic Press 1978)

3. **A brief course of higher mathematics:**
   ----
   by V.A.Kudryavtsev and B.P.Demidovich
   (Mir Publisher's Moscow 1980)

4. **Concice Encyclopedia of Mathematics:**
   ----
   by W.Gellert,H.Kustner,M Hellwich,H Kastner
   (Van Nostrand Reinhold Company New York and other cities 1977)

5. **Computational Mathematics:**
   ----
   by B.P Demitovich and I.A.Maron
   (Mir publishers Moscow 1976)

6. **Advanced calculus:**
   ----
   by Leopold Flatto
   (The Wiiliams and Wilkins Company - Baltimore 1982)

7. **Mathematics Handbook for Science and Engineering:**
   ----
                        ^
   by: Royal Lennart Rade and Bertil Westegren
   Fifth edition - 2004
   Springer Verlag Publications Inc
   Berlin - Heidelberg - New York

8. **Mathematical methods for physicists and engineers:**
   ----
   by: Royal Eugene Collins - 2nd corrected edition
   Dover Publications Inc - Mineola New York - USA 1991

9. **Differential Equations:**

   A systems approach - by: Jack Goldberg - and Merle C.Potter
   Prentice Hall International Editions
   Upper Saddle River , NJ - USA - 1998

- And a lot of personal work -

John.K.Bredakis MD

Assistant Professor University of Athens
American Board Certified Cardiologist

Born in Athens Greece 28/11/1946

**Graduate of the medical school (1970)
University of Athens**

**Trained in internal medicine and Cardiology
(1970-1977)
Chicago - USA**

**Consultant Cardiologist - Areteion Hospital Athens Greece
Since 1977**

Thanks God , uncle Fotis , Areteion Hospital
my parents , my wife Sofia
and professors C.Tountas , D.Voros , G.Limouris

A special thanks also to the professor Elias Kastanas
(Professor of mathematics - Engineer - Computer scientist etc)
The creator of my blog

I would also like to thank very much the professor of mathematics
Themistoklis Rassias

Athens Greece 2012

Pdf with remarks on Riemann Siegel formula
by Xavier Gourdon
Just to get an idea

1.3 The Riemann-Siegel formula

The *Riemann-Siegel formula* permits to approximate $\zeta(\sigma + it)$ in the critical strip $0 \leq \sigma \leq 1$ for large values of t with a number of terms proportional to $\sqrt{|t|}$ (which is much better than previous methods that requires a number of terms proportional to $|t|$). It states in the general form as follows.

Theorem 1 (Riemann-Siegel formula) *Let x and y be positive real numbers such that $2\pi xy = |t|$. Then for $s = \sigma + it$ with $0 \leq \sigma \leq 1$, we have*

$$\zeta(s) = \sum_{n \leq x} \frac{1}{n^s} + \chi(s) \sum_{n \leq y} \frac{1}{n^{1-s}} + O(x^{-\sigma}) + O(|t|^{1/2-\sigma} y^{\sigma-1}), \qquad (3)$$

where the O holds uniformly for $x > h$ and $y > h$ for any $h > 0$.

We remind that $\chi(s)$, often used to express functional equation of Zeta, is defined by

$$\chi(s) = 2^s \pi^{s-1} \sin\left(\frac{\pi s}{2}\right) \Gamma(1-s).$$

Formula (3) looks like the functional equation, for that reason it is also called the *approximate functional equation*. A proof can be found in [10] for example.

It is natural to use this formula when $x = y = \sqrt{|t|/(2\pi)}$, which writes in the form

$$\zeta(s) = \sum_{n=1}^{m} \frac{1}{n^s} + \chi(s) \sum_{n=1}^{m} \frac{1}{n^{1-s}} + E_m(s), \qquad m = \left\lfloor \left(\frac{|t|}{2\pi}\right)^{1/2} \right\rfloor \qquad (4)$$

where the error term $E_m(s)$ satisfies

$$E_m(s) = O(|t|^{-\sigma/2}).$$

An asymptotic expansion of the error term $E_m(s)$ can be given explicitly (see [10] or [8]), but it is complicated in its general form and we just give its first terms in the particular (but important) case $\sigma = 1/2$ below.

$$Z(t) = 2 \sum_{n=1}^{m} \frac{\cos(\theta(t) - t \log n)}{\sqrt{n}} +$$

$$(-1)^{m+1} \tau^{-1/2} \sum_{j=0}^{M} (-1)^j \tau^{-j} \Phi_j(z) + R_M(t)$$

$$Z(t) = 2\sum_{n=1}^{m} \frac{\cos(\theta(t) - t\log n)}{\sqrt{n}} +$$
$$(-1)^{m+1}\tau^{-1/2}\sum_{j=0}^{M}(-1)^j\tau^{-j}\Phi_j(z) + R_M(t)$$

This concise formula is very suitable for computation of $Z(t)$ for large values of t, as needed in computations relative to zeros of the zeta function. For precise computations, asymptotic formula of the error term is also available, and (using formulation of [2]) one has for positive t

$$Z(t) = 2\sum_{n=1}^{m} \frac{\cos(\theta(t) - t\log n)}{\sqrt{n}} +$$
$$(-1)^{m+1}\tau^{-1/2}\sum_{j=0}^{M}(-1)^j\tau^{-j}\Phi_j(z) + R_M(t), \tag{9}$$

with $R_M(t) = O(t^{-(2M+3)/4})$, where we used the notations

$$\tau = \sqrt{\frac{t}{2\pi}}, \quad m = \lfloor\tau\rfloor, \quad z = 2(t-m)-1.$$

The first functions $\Phi_j(z)$ are defined by

$$\Phi_0(z) = \frac{\cos(\frac{1}{2}\pi z^2 + \frac{3}{8}\pi)}{\cos(\pi z)}$$
$$\Phi_1(z) = \frac{1}{12\pi^2}\Phi_0^{(3)}(z)$$
$$\Phi_2(z) = \frac{1}{16\pi^2}\Phi_0^{(2)}(z) + \frac{1}{288\pi^4}\Phi_0^{(6)}(z)$$

The general expression of $\Phi_j(z)$ for $j > 2$ is quite complicated and we refer to [6] or [10] for it. As exposed in [2], explicit bounds have been rigorously obtained on the error term $R_M(t)$, and for $t \geq 200$, one has

$$|R_0(t)| \leq 0.127\,t^{-3/4}, \quad |R_1(t)| \leq 0.053\,t^{-5/4}, \quad |R_2(t)| \leq 0.011\,t^{-7/4}.$$

To be able to completly approximate $Z(t)$ thanks to formula (9), it remains to give an approximation of the $\theta(t)$ function which is obtained from expression (6) using Stirling series, giving

$$\theta(t) = \frac{t}{2}\log\frac{t}{2\pi} - \frac{t}{2} - \frac{\pi}{8} + \frac{1}{48t} + \frac{7}{5760t^3} + \cdots \tag{10}$$

Practical approximation considerations

For practical purposes in computations relative to zeros of $\zeta(s)$ it is not necessary to compute precisely the zeros but just to locate them, and using $M = 1$ or $M = 2$ in formula (9) is usually sufficient. For t around 10^{10} for example, the choice $M = 1$ permits to obtain an absolute precision of $Z(t)$ smaller than 2×10^{-14}, and with $M = 2$ the precision is smaller than 5×10^{-20}. As for the

number of terms involved in the sum of (9), it is proportional to $\sqrt{t}$ which is much better than previous approaches without Riemann-Siegel formula which required a number of terms of order t. For $t \simeq 10^{10}$ for example, Riemann-Siegel formula only requires $m \simeq 40,000$ terms, whereas approach of proposition 1 requires at least $\simeq 9 \times 10^9$ terms.

References

[1] D. Borwein, P. Borwein, A. Jakinovski, "An efficient algorithm for the Riemann Zeta function", (1995) available from the CECM preprint server, URL= http://www.cecm.sfu.ca/preprints/1995pp.html, CECM-95-043.

[2] J. Borwein, D. Bradley, R. Crandall, "Computational strategies for the Riemann Zeta Function", (1999) available from the CECM preprint server, URL= http://www.cecm.sfu.ca/preprints/1999pp.html, CECM-98-118.

[3] H. Cohen, F. Rodriguez Villegas, D. Zagier, *Convergence acceleration of alternating series*, Bonn, (1991)

[4] H. M. Edwards, *Riemann's Zeta Function*, Academic Press, 1974.

[5] J. Lagarias and A. Odlyzko, "Computing $\pi(x)$: an analytic method," *J. Algorithms* **8**, (1987), 173-191.

[6] Eric W. Weisstein, "Riemann-Siegel functions", available from the MathWorld site at URL= http://mathworld.wolfram.com/Riemann-SiegelFunctions.html

[7] A. Odlyzko, H. J. J. te Riele, "Disproof of the Mertens conjecture," *J. Reine angew. Math.* **357** (1985), 138-160.

[8] A. Odlyzko, A. Schönhage, "Fast algorithms for multiple evaluations of the Riemann zeta-function," *Trans. Amer. Math. Soc.* **309** (1988), 797-809.

[9] A. Odlyzko, "The 10^{20}-th zero of the Riemann zeta function and 175 million of its neighbors", unpublished manuscript, 1992. Available at URL= http://www.research.att.com/ amo/unpublished/index.html.

[10] E. C. Titchmarsh, *The theory of the Riemann Zeta-function*, Oxford Science publications, second edition, revised by D. R. Heath-Brown (1986).

[11] J. van de Lune, H. J. J. te Riele and D. T. Winter, On the zeros of the Riemann zeta function in the critical strip, IV. *Math. Comp.* **46** (1986), 667-681.